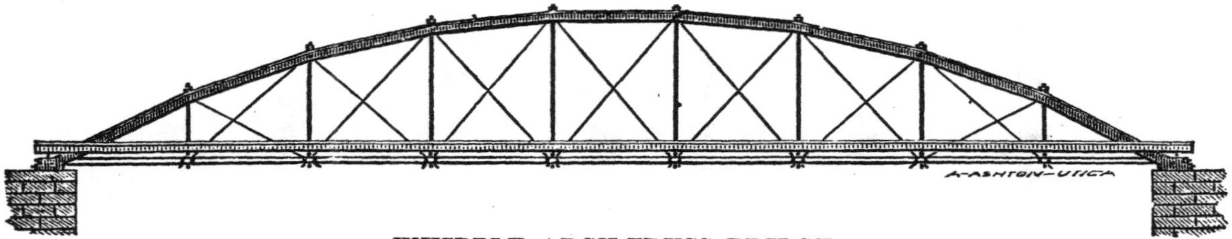

WHIPPLE ARCH-TRUSS BRIDGE.

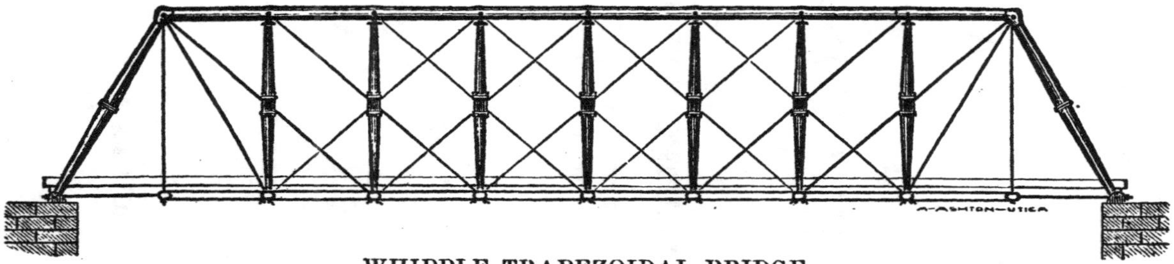

WHIPPLE TRAPEZOIDAL BRIDGE.

AN ELEMENTARY AND PRACTICAL TREATISE ON BRIDGE BUILDING,

AN

ENLARGED AND IMPROVED EDITION

OF THE AUTHOR'S ORIGINAL WORK,

BY

S. WHIPPLE, C.E.
ALBANY, N. Y.,
INVENTOR OF THE WHIPPLE BRIDGES, &C.

SECOND EDITION, REVISED AND ENLARGED.

NEW YORK:
D. VAN NOSTRAND, PUBLISHER,
23 Murray St., and 27 Warren St.
1873.

CONTENTS.

BRIDGE BUILDING.

PRELIMINARIES.

I. A bridge is a structure for sustaining the weight of carriages, animals, &c., during their transit over a stream, gulf or valley.

Bridges are constructed of various plans and dimensions, according to the circumstances and objects requiring their erection; and it is the purpose of this work, after a few remarks upon the general nature and principles of bridges, to attempt some analyses and comparisons of the respective qualities and merits of various general plans, with a view of deducing practical results, as to a judicious and economical choice and application of materials in the construction of these useful and important structures.

II. The force of gravity, on which the weight of bodies depends, acts in vertical lines, and consequently, a heavy body can only be prevented from falling to the earth, by a force equal and opposite to that with which gravity impels the body downward. This resisting force must not only act vertically upward, but the line of its action must pass through the centre of gravity of the body it sustains. All the forces in the world, acting parallel with, or perpendicular to, the vertical passing through its centre of gravity, could not prevent a musket ball (concentrated to the point of its centre of gravity) from falling to the centre of the earth, unless it were a horizontal force capable of giving the ball a projection, such that the centrifugal tendency should equal or exceed gravity—a kind of force which could never be made available toward preventing people from falling into the water in crossing rivers; consequently, having no application in bridge building.

In fact, nothing but a continuous series of unyielding material particles, extending from an elevated body downward to the earth, can hold or sustain that body above the earth, by vertical and horizontal action alone, either separately, or in combination.

III. Suppose a body, no matter how great or small, placed above the earth, with a deep void, or an inaccessible space beneath it. Attach as many cords to it as you please, strain them much or little — only horizontally — the body will fall, nevertheless. Thrust any number of rods, with whatever force you may, horizontally against it; still the body will fall. This is obvious from the fact that horizontal forces, acting at right angles with the direction of the force of gravity, have no more tendency to prevent, than to promote the fall of the body.

Moreover, the space beneath being inaccessible, there is no foundation, or foot hold, upon which to rest a post or stud that may directly resist the action of gravity, while the lines of all other vertical forces or resistances, pass by the body without touching it.

In the case here supposed, the body can only be prevented from falling by *oblique* forces; that is, by forces whose lines of action are neither exactly horizontal, nor exactly perpendicular. Attach two cords to the body, draw upon them obliquely upward and outward, in opposite directions, or from opposite sides of the void, with a certain stress, and the body will be sustained in its position. Apply two rods to it obliquely upward, of a proper degree of stiffness, in the same vertical plane, and on opposite sides of the perpendicular, a certain thrust exerted upon those rods, will prevent the descent of the body.

IV. Here, then, we have the elementary idea — the grand fundamental principle in bridge building. Whatever be the form of structure adopted, the elementary object to be accomplished is, to sustain a given weight in a given position, by a system of *oblique forces*, whose resultant shall pass through the centre of gravity of the body in a vertically upward direction, in circumstances where the weight can not be conveniently met by a simple force, in the same line with, and opposite to, that of gravity.

For a more clear illustration of this elementary idea, let us suppose *a a'*, Fig. 1, to represent the banks of a river, or the abutments of a bridge; and *gg'*, the line of transit for carriages, &c.; and, let us further suppose a load of a certain weight, *w*, to have arrived at a point centrally between *a a'*. The simplest method of sustaining the weight is, perhaps, either to erect two oblique braces *aw*. . *a'w*, or suspend two oblique chains or ties *pw*, *p'w*, from fixed supporting points *a a'*, or *p p'*.

It is not necessary that the weight be at the angular point *w*, of the braces or chains, but it may be sustained by simple suspension at *w'* below, or simple support at *w''* above, and such obliquity may be given to the braces or chains as may be most economical; a consideration which will be taken into account hereafter.

V. Thus we see how a weight may be sustained centrally between the banks of a river, or the extremities of a bridge. But the structure must not only provide for the support of weight at this point, but also at every other point between $a\,a'$, or $g\,g'$; and it is obvious that the same plan and arrangement will apply as well at any other point as at the centre, with only the variation of making the braces or chains of unequal length.

FIG. 1.

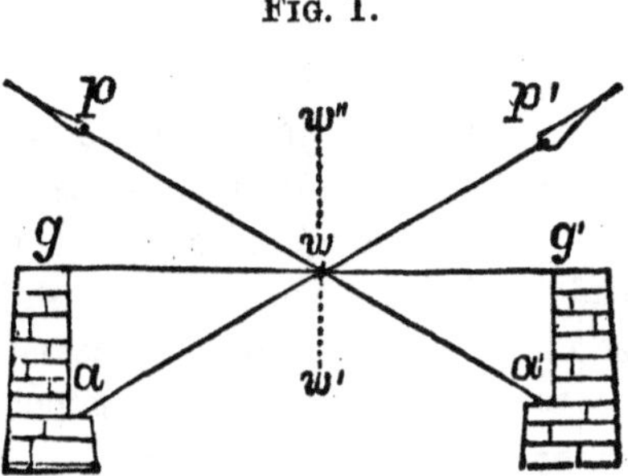

This, however, would require as many pairs of braces or chains as there were points between $g\,g'$, a thing, of course, impracticable, since the oblique members would interfere with one another, and be confounded into a solid mass. We therefore resort to the transverse strength and stiffness of beams,—phenomena with which all have more or less acquaintance, and without digressing in this place to investigate their principles and causes, it will be assumed as a fact sustained by all experience, that, for sustaining weight between two supporting points upon nearly the same level, a simple beam affords the most convenient and economical means, until those points exceed a certain distance asunder, which distance will vary with circumstances;

but in bridge building, will seldom be less than 10 to 14 feet, where timber beams are employed. Hence, for bridges of a length of 12 to 14 feet, usually, nothing better can be employed than a structure supported by longitudinal beams, with their ends resting upon abutments or supports upon the sides of the stream.

Of course, no reference is here had to stone or brick arches. For, though these are advantageously used for short spans, and in deep valleys, where the expense of constructing high abutments for supporting a lighter superstructure, would exceed or approximate to that of constructing the arch, it is the purpose of this work to speak only of those lighter structures, composed mostly of wood and iron, and supported by abutments and piers of stone, or by piles, or frames of wood.

Having then adopted the use of beams for supporting weight upon short spaces, it is only necessary upon longer stretches, to provide support for a point once in 10 or 14 feet, by braces, &c., from the extremities; and for intermediate points, to depend on beams or joists extending from one to another of the principal points provided for as above.*

VI. For a span of 20 or 30 feet, it would seem that no better plan could be devised, than to support a transverse beam midway between abutments, by two pairs of braces or suspension chains, proceeding from points at or over the abutments, one pair upon each side of the road-way; this transverse beam affording support for longitudinal beams or joists extending

* It is susceptible of easy demonstration that the power of beams to sustain weight by lateral stiffness, forms no exception to the principle that oblique forces alone can sustain heavy bodies over inaccessible spaces. But this matter is deferred for the present.

therefrom to the abutments. When suspension chains are used, it may properly be called a *suspension bridge.* If braces be employed, it is usually termed a *truss-bridge.*

HORIZONTAL ACTION OF OBLIQUE MEMBERS.

VII. Before advancing further, it will be proper to refer to an important principle or fact which has not yet been taken into account, though a fact by no means of secondary interest.

The sustaining of weight by oblique forces, gives rise to *horizontal* forces, for which it is necessary to provide counteraction and support, as well as for the weight of the structure and its load. The two equal and equally inclined braces, *ac* and *bc*, Fig. 2, in supporting the weight *w* at *c*, act in the directions of their respective lengths, each with a certain force, which is equivalent to the combined action of a vertical and a horizontal force, [*Elementary Mechanics—Statics,*] which may be called the vertical and horizontal *constituents* of the oblique force. These two constituent forces bear certain determinate relations to one another, and to the oblique force, depending upon the angle at which the oblique is inclined.

FIG. 2.

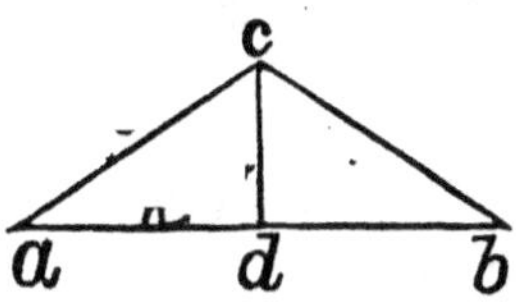

Now, we know that the vertical constituent alone contributes to the sustaining of the weight, and consequently, must be just equal to the weight sustained, in this case equal to $\frac{1}{2}w$. We know moreover, from the principles of statics, that three forces in equilibrio, must have their lines of action in the same plane, and meeting at one point; and must be respectively proportional to the sides of a triangle formed by lines drawn parallel with the directions of the three forces; and that each of the three forces is equal and opposite to the resultant of the combined action of the other two. We have, then, at *c*, the weight $\frac{1}{2}w$, the oblique force in the line *ac*, and a third force, equal and opposite to the horizontal constituent of the oblique force in the line *ac*. Then, letting fall the vertical *dc*, and drawing the horizontal *ad*, the sides of the triangle *acd*, are respectively parallel with the three forces in equilibrio at the point *c*. Hence, representing the vertical *cd*, by *v*, the horizontal *ad*, by *h*, and the oblique by *o*; and calling the horizontal force *x*, and the oblique force, *y*, we have the following proportions:

$$(1). \quad \tfrac{1}{2}w : x :: v : h, \text{ whence, } x = \tfrac{1}{2}w\frac{h}{v}$$

$$(2). \quad \tfrac{1}{2}w : y :: v : o, \text{ whence, } y = \tfrac{1}{2}w\frac{o}{v}$$

But $\frac{1}{2}w$ equals the weight sustained by the oblique *ac*. Therefore, from the two equations above deduced, we may enunciate the following important rule:

The horizontal thrust of an oblique brace, equals the weight sustained, multiplied by the horizontal and divided by the vertical reach of the brace; and the *direct* thrust (in the direction of its length), equals the weight sustained multiplied by the length, and divided by the vertical reach of the brace.

VIII. Now, it is obvious that the brace exerts the same action, both vertically and horizontally, at the lower, as at the upper end, though in the opposite directions; the brace being simply a medium for transmitting the action of weight from the upper to the lower end of the brace. Hence, the weight sustained by the brace *ac*, exerts the same vertical pressure at the point *a*, as it would do if resting at that point, while the brace requires a horizontal resistance to prevent its sliding to the left, as would be the case if its foot simply rested upon a smooth level surface. This horizontal resistance may be provided by abutments of such form, weight, and anchorage in the earth, as to enable them to resist horizontally as well as vertically, or by a horizontal tie, in the line *ab*, connecting the feet of opposite braces.

These two methods are both feasible to a certain extent, and in certain cases; and, both involve expense. Under particular circumstances, it may be a question whether the former should not be resorted to, wholly or partially. But for general practice, in the construction of bridges for heavy burthens, such as rail road bridges, and especially iron truss bridges, where expansion and contraction of materials produce considerable changes, it is undoubtedly best to provide means for withstanding the horizontal action of obliques, within the superstructure itself; and this principle will be adhered to in the discussions following.

The preceding remarks and illustrations as to the action of braces, or thrust obliques, obviously apply in like manner to obliques acting by tension, with only the distinction, that in the latter case, the weight is applied at the lower, and its action transmitted to the upper end of the oblique, and the horizontal action (at the remote end), is inward, and toward the vertical through the weight, instead of outward; and consequently, must be counteracted by outward thrust, as by a rigid body between the points *p p′*, Fig. 1, or by heavy towers, and anchorage capable of withstanding the inward tendency. Hence, in applying the rule before given, to *tension obliques*, and their vertical and horizontal constituents, the word *pull* should be substituted for the word *thrust*, wherever the latter occurs in said rule.

TWO PANEL TRUSSES.

IX. There are three forms of truss adaptable to bridges with a single central beam or cross bearer (which may be called two panel trusses), the general characteristics of which, are respectively represented by Figures 3, 4 and 5. Fig. 3 represents a pair of rafter braces, with feet connected by a horizontal tie, and with a vertical tie by which the beam is suspended at or near the horizontal tie, or the chord, as usually designated.

FIG. 3

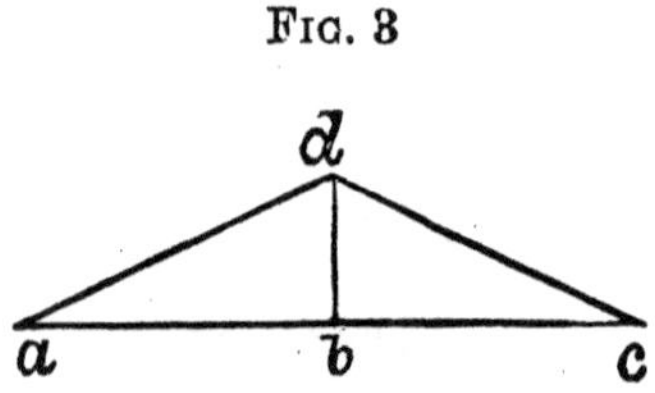

For convenience of comparison, let $bd = v, = 1, =$ vertical reach of oblique members in each figure. Also, let each chord equal $4v, = 4$, and the half chord $= 2 = h =$ horizontal reach of obliques in Figs. 3 and 4. Then *ad*, Fig. 3, equals $\sqrt{h^2+v^2} = \sqrt{5}$, and if the truss be loaded with a weight w, at the point b, bd will have a tension equal to w, and abc, [see rule at end of Sec. VII], a tension equal to $\frac{1}{2}w$, (= weight sustained by ad), multiplied by the horizontal, and divided by the vertical reach of ad; that is, equal to $\frac{1}{2}w\frac{h}{v}, = \frac{1}{2}w\frac{2}{1}, = w$; while ad suffers compression from end to end, equal to $\frac{1}{2}w\frac{ad}{v}$. But $ad = \sqrt{5}$, and $v = 1$. Whence $\frac{1}{2}w\frac{ad}{v} = \frac{1}{2}w\sqrt{5}$.

Now, as the cross-section of a piece, or member, exposed to tension (or to thrust, when pieces are similar in figure), should be as the stress, it follows that the *weight* of each such member should respectively, be as the stress sustained, multiplied by the length, + an additional amount taken up in forming connections; which latter, for purposes of comparing the general economy of different plans, may be neglected.

X. Then, representing by M, the amount of material required to sustain a stress equal to w, with a length equal to bd, = 1, we have only to multiply the stress of a member in terms of w, by the length in terms of bd, or v, and change w to M, to obtain the amount of material required for the member in question, omitting the extra amount in the connections. Hence, the length of the vertical tie bd, being equal to 1, and having a stress equal to w, requires an amount of material equal to 1M.

For the horizontal tie, or chord, length = 4, and stress (as seen above), = w, whence material = 4M. This added to 1M, required for the vertical, makes a total of 5M, for material exposed to tension in truss 3.

The two thrust braces, as already seen, sustain compression equal to $\frac{1}{2}w\sqrt{5}$, which multiplied by length, $\sqrt{5}$, and w changed to M, give material = $\frac{5}{2}$ M, for each, or 5 M, for the two.

XI. In the case of truss Fig. 4, the obliques manifestly sustain a weight = $\frac{1}{2}w$, by tension, giving stress = $\frac{1}{2}w\ \sqrt{5}$, which multiplied by length, = $\sqrt{5}$, gives $\frac{5}{2}$M = material for each, and 5M, for the two. The compression of ki, equals $\frac{1}{2}w \times h = \frac{1}{2}w \times 2 = w$, while the length = 4, whence, material = 4M; and, each end post sustaining $\frac{1}{2}w$, with length = 1, the two require material = M, making the whole amount of thrust material = 5M.

Thus we see that the two plans require each the same precise amount of material for sustaining both tension and thrust, upon the supposition that the material is capable of sustaining the same stress to the square inch of cross-section, in the one plan as in the other. This is true as to tension material; but with regard to thrust material, the power of withstanding compression, varies with the ratio of length to diameter of pieces, as well as with the form of cross-section; and it will hereafter be seen, that in this respect, plan 3 has the advantage in having the compression sustained mostly by shorter pieces, unless *ki* be supported vertically and laterally by a stiff connection with the beam at *f*, which would increase the amount of material.

FIG. 4.

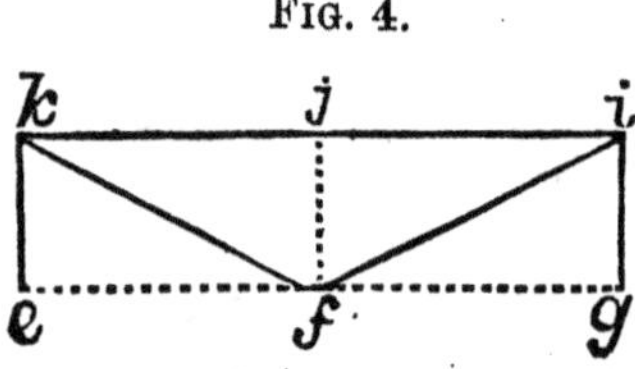

XII. Plan Fig. 5 has three members (*ln*, *mp* and *mq*) exposed to tension, and the remaining three exposed, to compression. With the same length and depth of truss, and the same load = w, at m, and with obliques equally inclined (at 45°), it is manifest that the vertical and horizontal reaches, each for each, is equal to 1, and the length, equal to $\sqrt{2}$; while the weight sustained by each equals $\frac{1}{2}w$. Hence, the action (of tension or compression), equals $\frac{1}{2}w\sqrt{2}$, and the material equals $\frac{1}{2}\sqrt{2}\times\sqrt{2}.\text{M} = 1\text{M}$;

FIG. 5.

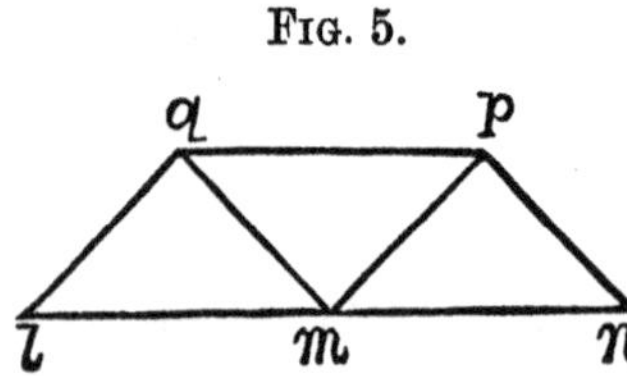

making for the four pieces, 2M for tension, and 2M for compression.

The tie or chord *ln*, suffers tension equal to the horizontal constituent of the thrust of *ql*, manifestly equal to the weight sustained by *ql*, or equal to $\frac{1}{2}$ *w*. Therefore, the length being equal to 4, the material required in its construction, equals 2M. The remaining member *pq* (= 2) sustains compression equal to the combined horizontal constituents of the tension of *mq*, and the compression of *ql*, each of said constituents equal to $\frac{1}{2}w$, making compression of *pq*, equal to *w*, and length being 2, material = 2M

We have therefore, for this plan of truss, 4M, for thrust material, and 4M for tension material, which is $\frac{1}{6}$ less than in case of Figs. 3 and 4. Consequently, this plan is decidedly more economical than either of the others, unless the compression material acts with better advantage in the latter than the former; that is, unless the thrust members in 3 and 4, have a greater power of resistance to the square inch of cross-section, than those in Fig. 5.

XIII. As to this, both theory and experiment prove, as will be shown in a subsequent part of this work, that the long thrust members in bridge trusses, are liable to be broken by deflection, rather than by a crushing of the material; that in pieces with similar cross-sections, with the same ratio of length to diameter, the power of resistance to the square inch is the same. That, since the cross-section is as the square of the diameter, and the diameters (in similar pieces), as the lengths, the *absolute* powers of resistance (being as the cross sections), are as the squares of the lengths.

Hence, if the compressive forces acting upon two pieces of different lengths, be to one another as the squares of the lengths of pieces respectively, and the diameters be as the lengths, the forces are as the cross-sections, and proportional to the power of resistance in each case, and the material in the two pieces, acts with equal advantage, as far as regards cross-section, so that the products of stress into length of pieces, are the true exponents of amount of material required in the two pieces respectively. It follows, that, if on dividing the forces acting upon the pieces in question respectively, by the squares of the lengths, the quotient be the same in both cases, the two pieces have the same power of resistance to the square inch, and in general, the greater the value of such quotient, the greater the power per inch, and the greater the economy, though not necessarily in the same précise ratio.

XIV. Applying this rule to thrust members in plan Fig. 3, being the braces, the compressive force equals $\frac{1}{2}w\sqrt{5}$, and square of length $= 5$. Hence the quotient $\frac{1}{10}w\sqrt{5} = 0.2236w$.

The piece *ki* Fig. 4, has length $= 4$ and compression $= w$, whence, force divided by square of length gives $\frac{1}{16}w = 0.0625w$. This shows the material to be capable of sustaining much more to the square inch in the former, than in the latter case, though it does not give the true ratio. On the other hand, *ek* and *gi*, with length $= 1$, and stress $= \frac{1}{2}w$, give quotient $= \frac{1}{2}w = 0.5w$. Hence, with similar cross-sections, these parts have greater power to the inch than either of the former, but not enough to balance the inferiority of *ki*, as compared with *ad* and *dc*, in Fig. 3.

With regard to truss Fig. 5, ql and pn, suffer each compression equal to $\frac{1}{2}w\sqrt{2}$, with square of length $= 2$, giving quotient $= \frac{1}{4}w\sqrt{2}, = 0.371w$, while pq, has compression $= w$, and square of length $= 4$, and quotient $= \frac{1}{4}w = 0.25w$. Hence it appears that this plan not only possesses a decided advantage in the less *amount of action** upon materials, but also, a considerable advantage as to ability of compression, or thrust members, to withstand the forces to which they are exposed.

XV. Still another modification for a truss to support a single beam, is formed by reversing Fig. 3, thus converting tension members into thrust members, and *vice versa;* the oblique members falling below, instead of rising above the grade, or road-way of the bridge. In this case, the long horizontal thrust member ac, is divided and supported in the centre, and its economy of action becomes the same as that of pq, in Fig. 5; and the truss gives the same exponents for both thrust and tension material as when in the position of Fig. 3. This arrangement affords no side protection, and is not always admissible, on account of interference with the necessary open space beneath.

DEDUCTIONS.

XVI. We seem to learn from what precedes, that:

(1). Since all heavy bodies not in motion toward, or not approaching the centre of the earth (or receding from it under the influence of previous impulse), exert a pressure equal to their respective weights [VIII],

* By the expression, *amount of action,* is meant, the sum of products of stresses into lengths of parts, or members.

either directly or indirectly upon the earth; and, since, a body crossing a bridge, having (as bridges are always supposed to have), a void space underneath, preventing a *direct* pressure, it follows, that every such body exerts an *in*direct pressure at some point or points at greater or less horizontal distance from the body.

(2). That the pressure of a body at a point or points not directly below it, can only take place through one or more intermediate bodies, or members, capable of exerting (by tension or thrust), one or more oblique forces upon the first named body, and it is the office of a bridge to furnish the medium of such horizontal transfer of pressure [IV].

(3. That a single oblique force can not alone prevent a heavy body from falling toward the earth (since two forces can only be in equilibrio when acting oppositely in the same line), and that each oblique force is equal to the combined action of a vertical and a horizontal constituent, of which the first alone is equal to the weight sustained and transferred by the oblique member, while the horizontal constituent, acting at both extremities of the oblique medium, must be counteracted by means outside of the oblique and the weight sustained by it; which means are usually to be supplied by other members of the structure [VIII].

(4). The direct force exerted by an oblique member (in the direction of its length), is equal to the weight sustained, multiplied by the length, and divided by the vertical reach of the oblique, while the horizontal constituent equals the weight sustained multiplied by the horizontal, and divided by the vertical reach of the oblique [VII].

(5). The amount of material required in a tension member, is as the stress multiplied by the length of the member [IX] (disregarding extras in connections), and the same is true of thrust members of similar formed cross-sections, sustaining stress proportional to the square of the length of pieces respectively.

(6). The respective stresses of two thrust members, divided by the squares of respective lengths, give quotients indicative of, though not proportional to, the relative efficiency of material in the two members, — the greater quotient showing the greater efficiency, or greater power of resistance to the square inch of cross-section [XIII].

With these rules or principles in view, we may proceed advantageously with general analyses and comparisons of different plans, or systems of bridge trussing, adapted to different lengths of span.

THREE PANEL TRUSSES.

XVII. In structures exceeding 25 or 30 feet in length, the length of joists from the centre to the ends, would require cross-sections so great, to give them the requisite stiffness, that their weight and cost would become objectionable. It becomes expedient, then, in such cases, to provide support for more than one principal point, or transverse beam, or bearer. A superstructure from 30 to 40 feet long, may be constructed with two cross beams, supported by two trusses with two pairs of braces each, with the feet connected by a horizontal tie or chord, as seen in Fig. 6.

The cross beams, may be at *b b′*, or suspended at *c* and *d*, at equal horizontal distances from *a a′*, and from one another; which latter position they will be regarded as occupying in this instance. Or, the figure may be inverted, thus reversing the action of the several thrust and tension members.

XVIII. Another, and a more common form of truss for two beams, is shown in Fig. 7. These may be called three panel trusses.

FIG. 6.

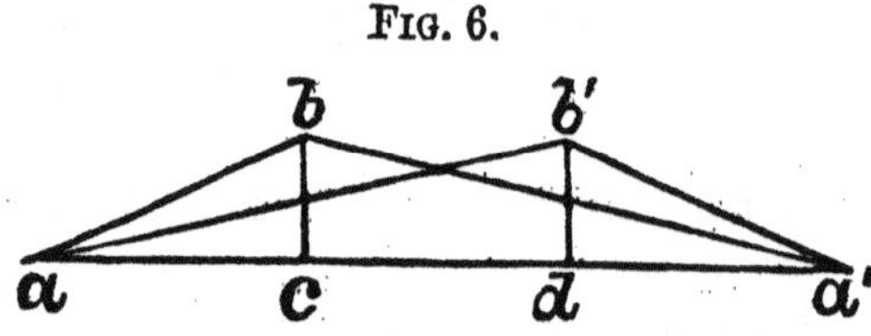

Fig. 7.

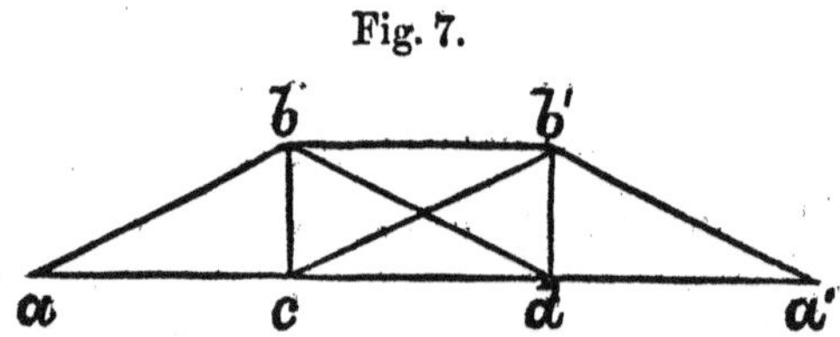

To compare these two trusses, suppose the two to have the same length and depth, and to be loaded with uniform weights, w, at the two points c and d in each. Then, since we know from the principle of the lever, that each weight produces upon each abutment, a pressure inversely as its horizontal distance from them respectively, and that the pressure upon the two abutments is equal to the weight producing it, it follows that ab, Fig. 6, sustains $\frac{2}{3}w$, and compression $= \frac{2}{3}\frac{ab}{bc}w$. Hence making $ab = D$,... $bc = v$, and $ac = h$, ... $\frac{2}{3}\frac{ab}{bc}w$, becomes $\frac{2}{3}\frac{D}{v}w$, and multiplying this stress by length, $= D$, and changing w to M*, we have for material in ab,... $\frac{2}{3}\frac{D^2}{v}$ M. But $D^2 = h^2 + v^2$, whence $\frac{2}{3}\frac{D^2}{v}M = \frac{2}{3}\left(\frac{h^2}{v} + v\right)M = \left(\frac{2h^2}{3v} + \frac{2v}{3}\right)M$.

Again, ab' sustains $\frac{1}{3}w$, with length $= \sqrt{4h^2 + v^2}$, and by multiplying and changing as in case of ab, we obtain material in ab', $= \left(\frac{4h^2}{3v} + \frac{v}{3}\right)$ M, which added to amount for ab, gives $\left(\frac{6h^2}{3v} + v\right)$ M for the two braces, and $\left(\frac{4h^2}{} + 2v\right)$M for the four.

The horizontal thrust of $ab = \frac{2}{3}w\frac{h}{v}$ while that of ab' $= \frac{1}{3}w\frac{2h}{v} = \frac{2}{3}w\frac{h}{v}$. Hence the horizontal thrust of ab and $ab = \frac{4}{3}w\frac{h}{v} =$ tension of aa', and material for chord aa', equals $3 \times \frac{4}{3}\frac{h^2}{v}M = \frac{4h^2}{v}M$. Tension of bc and $b'd$, each, equals w, and material for the two $= 2\,v$ M, which added to amount in aa', makes the whole tension material equal to $\left(\frac{4h^2}{v} + 2v\right)$ M, being the same co-efficient of M as was obtained for compression.

In truss Fig. 7, ... ab and $a'b'$ ($= D = \sqrt{h^2 + v^2}$), evidently sustain each a weight equal to w, and a stress $= \frac{\sqrt{h^2 + v^2}\,w}{v}$. Whence, material $= \left(\frac{h^2}{v} + v\right)$ M for each, and $\left(\frac{2h^2}{v} + 2v\right)$ M for both, while bb', equal to h, sustains compression equal to the horizontal thrust of ab, equal to $\frac{h}{v}w$, and requires material equal to $\frac{h^2}{v}$ M, making, with amount in braces ab $a'b'$, $\left(\frac{3h^2}{v} + 2v\right)$ M.

Now we have just seen that the horizontal thrust of ab, equal to the tension of chord aa', equals $\frac{h}{v}w$, and the

---

* When v is used in the co-efficient of M, then M represents the product of the stress, in terms of w, by length according to any assumed unit, which may be equal to v or not.

length being $3h$, the material consequently equals $\frac{3h^2}{v}$ M, to which add $2v$M for verticals, and we have $\left(\frac{3h^2}{v} + 2v\right)$M = whole amount of tension material in truss 7, which is less by $\frac{h^2}{v}$ M, than in the case of truss Fig. 6.

XIX. With regard to thrust material, it is clear that while in truss 6, the weight on either pair of braces, is transferred in due proportion to both abutments, independently of the other braces, whether one or both pair be loaded; on the contrary, truss 7, when c only is loaded, must transfer $\frac{1}{3}w$ to a', which can only be done through $a'b'$, the only oblique member acting at the point a'. Moreover, the weight must be communicated to $a'b'$, at the point b', through the thrust of bd, and the tension of db', assuming bd and cb' to be thrust members. Now, as either c, or d, is liable to be loaded with weight equal to w, while the other is unloaded, it follows that both bd and cb' are liable to sustain weight equal to $\frac{1}{3}w$, and require thrust material equal to $\frac{1}{3}\left(\frac{h^2}{v} + v\right)$ M for each, or $\left(\frac{2h^2}{3v} + \frac{2v}{3}\right)$M for the two. The whole amount of thrust material for truss 7, then, equals $\left(\frac{3h^2}{v} + 2v\right)$M, (the amount found above) + $\left(\frac{3h^2}{2v} + \frac{2}{3}v\right)$M, equal to $\left(\frac{3\frac{2}{3}h^2}{v} + 2\frac{2}{3}v\right)$M, against $\left(\frac{4h^2}{v} + 2v\right)$M for truss 6; the difference being $\left(\frac{\frac{1}{3}h^2}{v} - \frac{2}{3}v\right)$M. If this be a positive quantity, the balance is in favor of truss 7, and if negative, in favor of tress 6, as regards amount of action on thrust material; while, if $\frac{\frac{1}{3}h^2}{v} - \frac{2}{3}v$ = zero, the amount of thrust action is the same in both trusses.

Either of these suppositions may be true, according to the relative values of h and v. If $h = v\sqrt{2}$, then $\frac{\frac{1}{3}h^2}{v} - \frac{2}{3}v = o$. If h, be greater than $v\sqrt{2}$ .... $\frac{1}{3}\frac{h^2}{v} - \frac{2}{3}v$ is positive, and if h be less than $v\sqrt{2}$, the value is negative. But the amount is trifling in any probable relation of h and v, and may be disregarded in this general comparison.

Calling then, the *amount of action* upon thrust material in the two plans equal, there is a probable advantage in favor of truss 7, as to efficiency of thrust material, while the latter truss, shows a positive advantage over truss 6 in amount of tension material, equal to $\left(\frac{4h^2}{v} + 2v\right) - \left(\frac{3h^2}{v} + 2v\right)$ M $= \frac{h^2}{v}$ M. This is equal to 4M, when $h = 2v = 2$; which is in tolerable proportion for the trusses under discussion, and, substituting these values of h and v in the expressions of tension material in the trusses respectively, we have (16 + 2)M for truss 6, and (12 + 2)M for truss 7, being about 28½ per cent more for 6 than for 7.

The same difference would appear with Fig. 6 inverted, the thrust and tension action being the same in amount of each, only sustained by different members, thrust members in one case becoming tension members in the other.

FIVE PANEL TRUSSES.

XX. Truss Fig. 7, may be increased in length and number of panels, by introducing additional panels between the end triangular panels, and the rectangular centre one either of an oblique form, as in Fig. 8, which represents an arch-truss, or of a rectangular form as in

Fig. 10. The truss on the plan of Fig. 6, may be lengthened by introducing additional pairs of independent braces, as seen in Fig. 9.

FIG. 8.

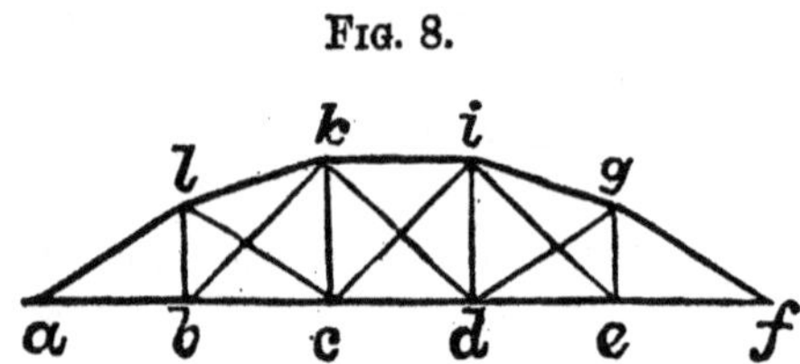

For the analysis of these trusses, using the same notation as before, as far as applicable, that is, making v = verticals ck, in 8 and 10, and equal to nn', pm', etc., in Fig. 9; $h = ab$,= width of panel in each figure, = $\frac{1}{5}$ whole chord; w=uniform weights at the four bearing points in each, and M = weight of material required to sustain a stress equal to w, with length equal to 1; then, making $lb = \frac{2}{3}v$ in truss 8, it is obvious that the two abutments at a and f together sustain $4w$, with the common centre of gravity of all the weights midway between abutments, whence each abutment sustains $2w$, equal to weight sustained by al. The compression of al therefore equals $\frac{2al}{\frac{2}{3}v}w$. But $al = \sqrt{h^2 + \frac{4}{9}v^2}$, which substituted in last preceding expression, gives $\frac{6}{2}\frac{\sqrt{h^2 + \frac{4}{9}v^2}}{v}w$,=compression of al. Whence, multiplying by length, $\sqrt{h^2+\frac{4}{9}v^2}$, and changing w to M, we have $\left(\frac{3h^2}{v}+1\frac{1}{3}v\right)$ M = material for al.

The horizontal thrust of al, [XVI (4)] equals $2w\frac{h}{\frac{2}{3}v} = \frac{3h}{v}w$, = tension of chord af.

The oblique member lk, sustains weight=w (through the vertical ck), and has a vertical reach = $\frac{1}{3}v$, whence it suffers compression equal to $\frac{lk}{\frac{1}{3}v}w$, $=\frac{3lk}{v}w$, $= 3\frac{\sqrt{h^2+\frac{1}{9}v^2}}{v}w$, and requires material equal to $\left(\frac{3h^2}{v}+\frac{1}{3}v\right)$M, while its horizontal thrust equals $\frac{h}{\frac{1}{3}v}w$, $=\frac{3h}{v}w$, $=$ compression of *ki*, by which it is contracted. The material required for *ki*, therefore, $=\frac{3h^2}{v}$M. Material for *ig* and *gf*, is the same as above found for *al* and *lk*, and, doubling those quantities, and adding amount just found for *ki*, we obtain $\left(\frac{15h^2}{v}+3\frac{1}{3}v\right)$ M, $=$ material in the whole arch.

The tension of the chord *af* (Fig. 8), has been seen to be equal to $\frac{3h}{v}w$, whence, multiplying by the length, $5h$, and changing w to M, we have $\frac{15h^2}{v}$M, $=$ material for chord.

The 4 verticals sustain each, weight $= w$, and the aggregate length being $3\frac{1}{3}v$,... material $= 3\frac{1}{3}v$M. This, added to amount in chord, gives $\left(\frac{15h^2}{v}+3\frac{1}{3}v\right)$M, $=$ tension material required to support a full uniform load, as above assumed. But since any number of the points *b*, *c*, *d*, *e*, are liable to be loaded while the others are unloaded, it is obvious that in such case, the arch will not be in equilibrio, the loaded points tending to be depressed, while the unloaded, tend to be thrust upward. Hence the arch requires the action of the obliques, or diagonals, in the three quadrangular panels, to counteract such tendency; and, as will appear further on, these members will require material equal to about one-third of the amount required in the chord, thus increasing the amount of tension material for the truss to about $\left(\frac{20h^2}{v}+4\frac{1}{3}v\right)$M.

XXI. In truss Fig. 9, each brace obviously sustains a portion, x, of the weight w, which is to w, as the horizontal reach of its antagonist, as to horizontal action, or its fellow and assistant vertically, is to the whole length of chord; that is, the weight x, bearing at m, through mn', is to w, as ns to ms; or, $x : w :: 4 : 5$. Hence $x = \frac{4}{5}w$. This, multiplied by the horizontal reach, equal to h, and divided by v, gives the horizontal thrust of the brace, equal to $\frac{4}{5}\frac{h}{v}w$.

FIG. 9.

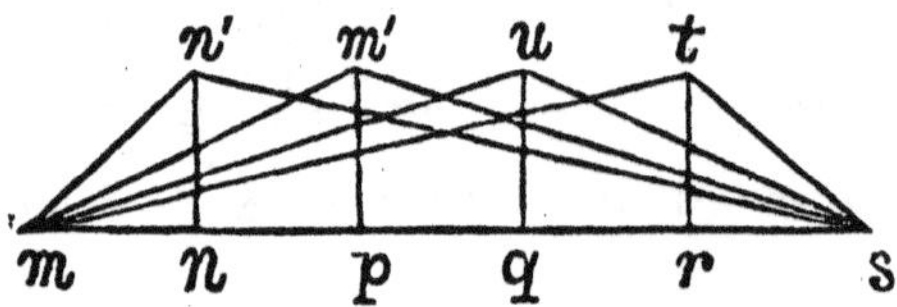

In like manner, mm' sustains a weight x', which is to w, as ps to ms, i. e., $x' : w :: 3 : 5$, whence $x' = \frac{3}{5}w$, and the horizontal thrust $= \frac{3}{5}\frac{2h}{v}w = \frac{6}{5}\frac{h}{v}w$; and in general, the horizontal thrust of a brace in this kind of truss, equals w, multiplied by the product of the number of sections of chord at the right and the left of the point of application (of the weight), and divided by the whole number of chord-sections, and, by the vertical reach (v) of the brace.

But the horizontal thrust of mn' equals that of $n's$, = that of mt, and the horizontal thrust of mm', equals that of $m's$ = that of mu, whence, horizontal thrust of the 4 braces bearing at m, equals twice that of mn' and mm', together, $= 2\left(\frac{4}{5}\frac{h}{v} + \frac{6}{5}\frac{h}{v}\right)w = \frac{20h}{5v}w = 4\frac{h}{v}w$. This being equal to the tension of the chord, multiplying by length $5h$, and changing w to M, gives material for chord $= \frac{20h^2}{v}$M. Adding to this, $4v$M, for 4 verticals with stress $= w$, and length $= v$, each, makes whole amount of tension material equal to $\left(\frac{20h^2}{v}+4v\right)$M, being very nearly the same as for truss, Fig. 8.

XXII. As for braces in truss 9, we have already seen that each brace sustains weight equal to w, multitiplied by the number of panels crossed by its fellow, and divided by the number of panels in the whole truss. Hence mn' sustains $\frac{4}{5}w$, with length equal to $\sqrt{h^2+v^2}$. Therefore, stress $= \frac{4}{5}\frac{\sqrt{h^2+v^2}\,w}{v}$, which multiplied by length, and w changed to M, gives material for $mn' = \frac{4}{5}\left(\frac{h^2}{v}+v\right)$M, $= \left(\frac{4h^2}{5v}+\frac{4v}{5}\right)$M. mm' sustains $\frac{3}{5}w$ with length $= \sqrt{4h^2+v^2}$, whence material $= \frac{3}{5}\left(\frac{4h^2}{v}+v\right)$M, $= \left(\frac{12h^2}{5v}+\frac{3v}{5}\right)$M. mu sustains $\frac{2}{5}w$ with length $= \sqrt{9h^2+v^2}$, and material equals $\left(\frac{18h^2}{5v}+\frac{2v}{5}\right)$M, while mt, sustains $\frac{1}{5}w$, with length $= \sqrt{16h^2+v^2}$ requiring material $= \left(\frac{16h^2}{5v}+\frac{v}{5}\right)$M. Then, adding and doubling these amounts, we obtain $\left(\frac{20h^2}{v}+4v\right)$M, against $\left(\frac{15h^2}{v}+3\frac{1}{5}v\right)$M, for truss 8; a difference of about 30.6 per cent when $h = v$, and about 32.6 per cent when $h = 2v$.

Thus, truss Fig. 8 has over 30 per cent advantage over truss Fig. 9, in the economy of *amount* of *action* upon thrust material, with the advantage as to *efficiency* of action of this material, undoubtedly, also on the side of truss 8. Tension material is nearly the same in both.

XXIII. If truss Fig. 9 be inverted, dropping the oblique members below the road-way of the bridge, thus reversing the action of thrust and tension members, the thrust material would act with nearly equal advantage in both plans, and with about the same amount of action. But the 30 per cent advantage as to amount of action upon *tension* material, would still be in favor truss 8. Besides, it is only in exceptional cases that this arrangement can be adopted, on account of interference with the necessary space below the bridge.

FIG. 10.

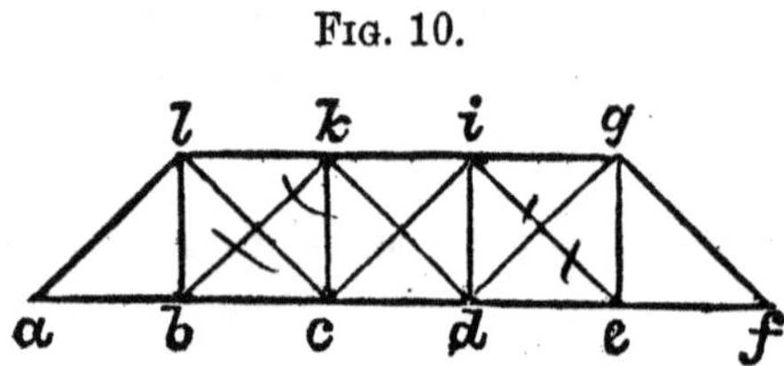

XXIV. In truss Fig. 10, suppose the points *b*, *c*, *d*, *e*, to be loaded successively from left to right, with uniform weights equal to *w* each, and suppose the truss to be without weight, as we have hitherto done. When *b* alone is loaded, $\frac{1}{5}w$ must bear at *f*, [XVIII] which may be effected, either by tension of *bl*, thrust of *lc*, tension of *ck*, thrust of *kd*, &c., by tension vertical and thrust diagonal alternately, till it reaches *f*; acting in its course upon 4 verticals, and 4 obliques, with a weight upon each, equal to $\frac{1}{5}w$. Or, the weight may be transferred by tension of *bk*, *ci* and *dg*, and thrust of *kc*, *id* and *gf*. These alternatives are subject to the control of the builder, and he will form and connect the parts accordingly. Let it be assumed that the truss has tension diagonals, and thrust uprights at *c* and *d*, while *lb*

and *eg* are necessarily tension members in all cases, in practice.

Again a weight, *w*, at *c*, must cause pressure equal to $\frac{2}{5}w$ at *f*, through tension of *ci* and *dg*, and thrust of *id* and *gf*. This, with the $\frac{1}{5}w$, from the weight at *b*, makes $\frac{3}{5}w$, acting on *ci*. But the weight at *c*, causes pressure equal to $\frac{3}{5}w$ at *a*, necessarily through tension of *cl*; and since *cl* and *bk* are antagonistic, the action upon one tending to produce relaxation of the other, it follows that only one can act at the same time, unless unduly strained in the adjustment of the truss. Hence, the $\frac{1}{5}w$, which acts upon *bk*, when *b* alone is loaded, is overbalanced by the $\frac{3}{5}$ *w*, tending to act upon *cl*, on account of the load at *c*; and the result is, that *bk* is relaxed, the whole weight at *b*, is necessarily sustained by *bl* and *la*, and the $\frac{3}{5}w$, which must by a statical necessity, bear at *f*, in consequence of the loads at *b* and *c*, is all made up from the weight at *c*, leaving only $\frac{2}{5}w$ of this weight to bear at *a*, through *cl*. Now, since it is obvious that all load at *c*, *d*, or *e*, must contribute to the pressure at *a*, which can only occur through action upon *cl*, it follows that *bk* can only sustain the whole weight of $\frac{1}{5}w$, when the point *b* alone is loaded; and consequently, that $\frac{1}{5}w$ is the greatest weight that *bk* can ever be subjected to.

Then, applying another weight, *w*, at *d*, it must add $\frac{3}{5}w$ to the pressure at *f*, through tension of *dg* and thrust of *gf*; which last amount, added to $\frac{3}{5}w$, communicated to *dg* through *ci* and *id*, makes $\frac{6}{5}w$, as the weight sustained by *dg*. But the weight at *d*, also causes pressure at *a*, equal to $\frac{2}{5}w$, which can only be done through action, or tendency to action upon *dk*, and since *dk* and *ci* are antagonists, only one can act at once, and that, only with a force equal to the excess of tendency to action of the one, over that of the other. Now we have seen that weights at *b* and *c*, tend to throw $\frac{3}{5}w$ upon *ci*, while the weight at *d*, tends to throw $\frac{2}{5}w$ upon *dk*. Hence, in these circumstances, *ci* only sustains $\frac{1}{5}w$, which is transferred to *dg* through thrust of *id*, while *dk* is relaxed, and the whole weight at *d*, is sustained by *dg*; making, with the $\frac{1}{5}w$ from *ci*, just above mentioned, $\frac{6}{5}w$, equal to the pressure due upon the abutment at *f*, on account of weights at *b*, *c* and *d*.

Lastly, a weight, *w*, at *e*, tends to give pressure equal to $\frac{4}{5}w$ at *f*, through *eg* and *gf*, and a pressure equal to $\frac{1}{5}$ *w* at *a*, through *ei*, *dk*, etc. This latter tendency has the effect to diminish by $\frac{1}{5}w$, the tendency of previously imposed weights, to throw $\frac{6}{5}w$ upon *dg*, reducing it to $\frac{5}{5}w$, and to neutralize the balance of $\frac{1}{5}w$ acting upon *ci*, after the imposition of the weight at *d*, leaving *ci* and *dk* both inactive, while *eg* sustains the whole weight applied at *e*, equal to *w*.

Now, as we have seen, any weight at *d* or *e*, tends to throw action upon *dk*, thereby diminishing action upon *ci*, and since weight at *b* and *c*, both contribute to the stress of *ci*, it follows that the maximum action upon *ci*, occurs when *b* and *c* are loaded, and *d* and *e*, unloaded.

For similar reasons, the maximum action upon *dg*, occurs when *e* alone is unloaded.

The maximum weight sustained by *lb*, and *eg*, is the weight applied directly at each of the points *b* and *e*, equal to *w*, and the maximum weights sustained by *ei*, *dk*, and *cl*, are the same as those sustained by *bk*, *ci* and *dg*, each respectively, as just above determined; while *al* and *gf*, both receiving action from weight on any part of the truss, obviously sustain their maximum weight, equal to $2w$, under the full load of the truss.

The section *ab*, of the lower chord, suffers a stress equal to the horizontal thrust of *al*, which of course, is greatest when *al* sustains the greatest weight. This has just been seen to be equal to $2w$, and occurs under a full load of the truss. Hence the greatest stress upon *ab* equals $2w\frac{h}{v}$, and is communicated without change to *bc*, *bk* being inactive when the truss is fully loaded. The section *cd*, suffers stress equal to the combined horizontal action of *al* and *lc*, which must be greatest when this combined action is greatest. That is also under the full load of the truss. For, though *lc* sustains $\frac{1}{5}w$ more weight when *b* is unloaded, the same cause relieves *al* of the amount of $\frac{4}{5}w$. Consequently, the weight borne by the two, is $\frac{3}{5}w$ less in this case, than when the truss is fully loaded. The greatest combined weights, then sustained by *al* and *lc*, being equal to $3w$, the greatest stress of *cd* equals $3w\frac{h}{v}$. This is also the greatest compression suffered by the upper chord *lg*, since the latter is also equal to the combined horizontal thrust and pull of *al* and *lc*. The stress of this chord is the same throughout, because the obliques meeting at *k* and *i*, are inactive when the truss is loaded throughout.

The maximum compression upon *ck* and *id*, equals the greatest weight sustained by *ci* and *dk*, already found to be equal to $\frac{3}{5}w$.

XXV. Having thus ascertained the greatest weights sustained by the several oblique members, and the greatest stresses of the horizontals and verticals, we may deduce the required amount of material, or, perhaps more properly, the *amount of action* upon the material required for the truss, as compared with like amount of action in trusses 8 and 9, thus: Max. weight on end braces, $2w \times$ length $\sqrt{h^2 + v^2}$ .... stress $= \frac{2w\sqrt{h^2 + v^2}}{v}$.

Hence, action upon material for the two = ...	$\left(\frac{4h^2}{v} + 4v\right)$M.
Max. weight on 2 verticals $= \frac{3}{5}w \times$ length of the two ($= 2v$), gives......	$1\frac{1}{5}v$ M.
Max. stress of upper chord $= 3w\frac{h}{v}$ $\times$ length ($= 3h$), gives amount of action = ...	$9\frac{h^2}{v}$ M.
Making total amount of action on thrust material = ...	$\left(\frac{13h^2}{v} + 5\frac{1}{5}v\right)$ M
Aggregate max. weight on 6 tension diagonals $= \frac{20}{5}w = 4w$. This by the length ($= \sqrt{h^2 + v^2}$), gives stress $= 4\,w\frac{\sqrt{h^2 + v^2}}{v}$; whence amount of action on material, equals	$\left(\frac{4h^2}{v} + 4v\right)$M.
2 tension verticals sustain each, $1w$, with length $= v$, giving amount of action for the two = ...	$2v$ M.
Stress of middle section, lower chord $= 3w\frac{h}{v}$, $\times$ length ($= h$), gives action	$8\frac{h^2}{v}$M.
4 remaining sections, with stress $= 2w\frac{h}{v} \times$ length ($= 4h$), give	$8\frac{h^2}{v}$M.
Making whole amount of action on tension material = ...	$\left(15\frac{h^2}{v} + 6v\right)$M.

SYNOPTICAL STATEMENT IN REGARD TO TRUSSES (Figs. 8, 9 and 10.

No. of Plan.	References.	Amount of Action upon Materials		
		Tension.	*Compression.*	*Total.*
8	XX	$\left(\frac{20h^2}{v}+4\frac{1}{3}v\right)$ M	$\left(\frac{15h^2}{v}+3\frac{1}{3}v\right)$ M	$\left(\frac{35h^2}{v}+7\frac{2}{3}v\right)$ M
9	XXI, XXII	$\left(\frac{20h^2}{v}+4v\right)$ M	$\left(\frac{20h^2}{v}+4v\right)$ M	$\left(\frac{40h^2}{v}+8v\right)$ M
10		$\left(\frac{15h^2}{v}+6v\right)$ M	$\left(\frac{13h^2}{v}+5\frac{1}{5}v\right)$ M	$\left(\frac{28h^2}{v}+11\frac{1}{5}v\right)$ M

Making $h=v=1$, the above table will be as follows:

8 .... $24\frac{1}{3}$M .... $18\frac{1}{3}$M ....$42\frac{2}{3}$M
9 .... 24M............ 24M 48M
10 .... 21M............ $18\frac{1}{5}$M ....$39\frac{1}{5}$M

XXVI. This shows very nearly the relative amount of tension material required in the several plans; while, as previously stated, the amount of compression material is not so nearly indicated by the figures and expressions giving the *amount of action* (sum of stresses into lengths of pieces), as in case of tension members. The compression material in No. 8 (the arch truss), is undoubtedly more efficient in action than in either of the others, while that in No. 9, is unquestionably the least so. In fact, this truss will be hardly considered as possessing advantages of any kind, sufficient to induce its adoption; and it will not be considered in the discussions and comparisons in regard to trusses of greater span, to which we may now proceed.

TRUSSES WITH SEVEN PANELS.

THE ARCH TRUSS.

XXVII. In Fig. 11, let $md = v$, and $h = ab = \frac{ai}{7}$. If each of the points b, c, d, etc., be loaded with a weight equal to w, then, in order that the arch may be in equilibrio under the effects of these weights, without any action of the diagonals, it is necessary that each section of the arch have the same horizontal thrust, since, if one section have a greater horizontal thrust than the one opposed to it at either end, the diagonals alone can sustain the surplus. And, that the sections may have the same horizontal thrust each must have a vertical reach (the horizontal reach being the same for all), proportional to the weight (W) sustained by each. For illustration, horizontal thrust being equal to $W\frac{h}{v}$, in order that this expression may represent a constant quantity, h remaining constant, v must be as W.

Now, ml being horizontal, can sustain none of the weight acting at the point m, through the vertical md; hence mn must sustain a weight equal to w. This is transferred to no [VIII], and in addition to the weight at c, makes $2w$ sustained by no. The latter weight is in turn transferred to oa, and, in addition to the weight at b, makes $3w$, to be sustained by ao. The vertical reaches, therefore, beginning with ao, should be as 3, 2 and 1; whence, ob should equal $\frac{1}{2}v$, and nc should equal $\frac{5}{6}v$.

The thrust of ao, then, equals $3w\frac{ao}{\frac{1}{2}v}$, $= 6w\frac{ao}{v}$, $= 6w\frac{\sqrt{h^2+\frac{1}{4}v^2}}{v}$; whence, amount of action on material* $= \ldots\ldots\ldots\ldots\left(\frac{6h^2}{v} + 1\frac{1}{2}v\right)\times$ M.

Thrust of on, $= 2w\frac{on}{\frac{2}{3}v}$, $= 6w\frac{on}{v}$, $= 6w\frac{\sqrt{h^2+\frac{1}{9}v^2}}{v}$, and material $= \ldots\ldots\ldots\ldots\left(\frac{6h^2}{v} + \frac{2}{3}v\right)\times$ M.

Thrust of $nm = w\frac{nm}{\frac{1}{6}v} = 6w\frac{nm}{v} = 6w\frac{\sqrt{h^2+\frac{1}{36}v^2}}{v}$, and material $= \ldots\ldots\ldots\ldots\left(\frac{6h^2}{v} + \frac{1}{6}v\right)\times$ M.

Thrust of ml (= horizontal thrust of nm), $= w\frac{h}{\frac{1}{6}v} = 6w\frac{h}{v}$, and material $= \ldots\ldots\ldots\ldots 6\frac{h^2}{v}\times$ M.

Adding these amounts, and repeating the first three, we have $\left(\frac{42h^2}{v} + 4\frac{2}{3}v\right)\times$ M, equal to amount of action upon the arch when fully loaded.

FIG. 11.

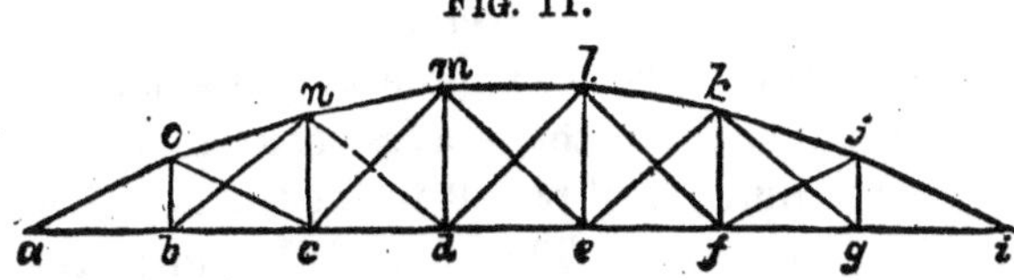

The stress of the chord obviously equals the horizontal thrust of ao, equal to $3w\frac{h}{\frac{1}{2}v}$, $= 6w\frac{h}{v}$; and is the same throughout, when the truss is fully loaded throughout. Hence, for the whole chord, we have, stress $= 6w\frac{h}{v}$ multiplied by length ($= 7h$), and w changed to M, $= 42\frac{h^2}{v}\times$ M, representing the material required for the chord.

The above are assumed, for the present, to be the greatest stresses that any part of the chord or arch can

* By amount of action upon material, is meant the stress of a member multiplied by its length.

be subjected to, in any condition of the load; w, being the maximum weight for any one of the sustaining points, b, c, d, &c. This is a point we shall be better enabled to verify after considering the

STRESSES OF VERTICALS AND DIAGONALS.

XXVIII. As the diagonals do not act under a full load of the truss, the verticals must each sustain a tension equal to w, when the weights are applied at the chord; and, the diagonals acting by tension, serve, when in action, to diminish the tension of verticals, or to subject them to compression, but can never *increase* their action of tension. Hence, the maximum tension stress of each vertical equals w.

In order to bring the diagonals into action, the truss must obviously be unequally loaded; and, to determine the *maximum* stresses of the several diagonals respectively, we may begin by removing the weight at g; (see Fig. 11A), and, to facilitate the process, let w'' represent w divided by the number of panels in the truss (= 7 in this case), i. e., $w'' = \frac{1}{7}w$. Then, the full load of the truss bearing with a weight of $3w$, $= 21w''$, at i, ... with load removed from g, the bearing at i, equals $21w'' - 6w''$, $= 15w''$; and produces a thrust upon ij equal to $15w''\frac{\sqrt{h^2+\frac{1}{4}v^2}}{\frac{1}{2}v}$. Then, taking jq by any convenient scale, on ij produced, to represent the thrust of ij (reduced to w'' with a numerical coefficient, according to the proportions of the truss), and drawing qr parallel with fj, and meeting jk produced in r, it is obvious that the three forces acting at j, namely, the thrust of ij and jk, and the tension of fj, will be represented respectively by the sides of the triangle jqr, parallel respectively with the directions of those forces; and may be measured by scale and dividers, or calculated trigonometrically.

Now, it will be seen that the greater the pressure at *i*, the greater the thrust of *ij*, represented by *jq*, and consequently, the greater the line *qr*, representing the tension of *fj*. But the pressure at *i*, is manifestly the greatest possible in the case here supposed, except when the weight at *g* is wholly or partially restored, in which case the tension of *fj* would be wholly or partially relieved. Hence, it follows that the maximum stress of *fj*, occurs when all the supporting points except *g*, have their full load, and the point *g* is without load.

Taking, then, $fs = qr$, and drawing *st* parallel with *gf*, *st* will represent the horizontal, and *ft*, the vertical effect (equal to $3w''$) of the action of *fj*; the former

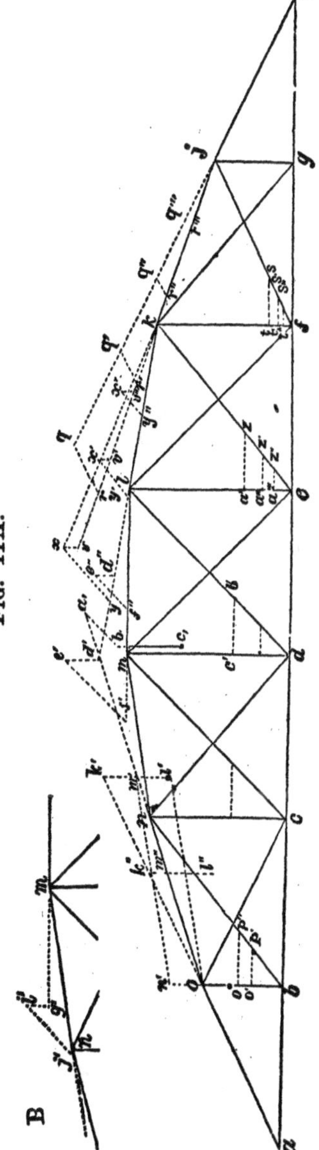

FIG. 11A.

www.ingramcontent.com/pod-product-compliance
Lightning Source LLC
LaVergne TN
LVHW021252110826
845151LV00005BA/1499